Filiusventi

RESEARCH ON THE KAMAL
including
KNOT-SPACING ALGORITHMS
and a theory about
THE PRECISION TRICK

Wndsn Applied Science Lab
Research Dispatch

Tracing the research and the origin of an ancient instrument, including a comparison of various methods to derive knot-spacing, as well as original research on how to increase reach and precision with a limited size instrument. Also computing the optimum size of a Kamal for a given set of altitudes.

BERLIN:
Published by the Wndsn applied science lab.
M.M.XVIII.

Filiusventi
Research on the Kamal

The Knot-spacing Algorithms used,
and a Theory about the Precision Trick

Published by the Wndsn applied science lab
© Copyright 2018 Wndsn XPD

Arcane science for sophisticated adventurers
www.wndsn.com

This book was typeset using the LaTeX document processing system originally developed by Leslie Lamport, based on the TeX typesetting system created by Donald Knuth.

Contents

Description of the Instrument   1
    What is a Kamal? 1
    General Construction 3
    Knot-Spacing Algorithms 4
    The Cross-Staff, a Close Relative 6

Reconstructing the Kamal   9
    Data 9
    Questions 9
    Size and Proportion 10

A Theory on Re-using Knots to Increase Precision   13
    The Precision Trick 13
    Two Strings? 16
    Calculating Optimum Kamal Size 18
    Conclusion 19

Precision, Accuracy, and Range   21
    Achieved Precision 21
    Achieved Range 22
    Accuracy 22

Bibliography   24

About Wndsn   25

List of Figures

1	Principle of Kamal.	2
2	Kamal reconstruction.	3
3	Kamal vs. cross-staff graduation.	7
4	Two strings?	17

List of Tables

1	Comparison of different knot intervals.	8
2	Sample degree intervals.	11
3	Corresponding values across scales and sides.	14
4	Demonstrating scale re-use for 1° to 0.5°.	15
5	Demonstrating scale re-use for 2° to 1°.	16
6	Knots for optimum Kamal dimensions.	20

Description of the Instrument

What is a Kamal?

Purpose

The Kamal is a historical instrument mainly used in seafaring for celestial navigation and latitude sailing, knowing the latitude of a destination and sailing against that until making landfall.

> [...] an instrument for taking the altitude of polar and circumpolar stars in its most elementary shape. It consists of a small parallelogram of horn (about two inches by one) with a string (or a couple of strings, in some instances), inserted in the center. On the string are nine knots. To use the instrument for taking the height of Polaris, the string is held between the teeth, with the horn at such distance from the eye, that while the lower edge seems to touch the oceanic horizon, the upper edge just meets the star: the division or knot is then read off as the required latitude. [Prinsep, 1928]

The Kamal was used to determine angles, for instance the angle between the horizon and Polaris or the sun to determine a vessel's latitude, or the angle between the top and bottom of an object to determine the distance to said object if its height is known, or the height of the object if its distance is known, or the horizontal angle between two visible locations to determine one's point on a map. The Kamal has a certain similarity to the cross-staff, where the string is replaced by a stick, and the general principle remains the same. (See figure 1 on page 2.)

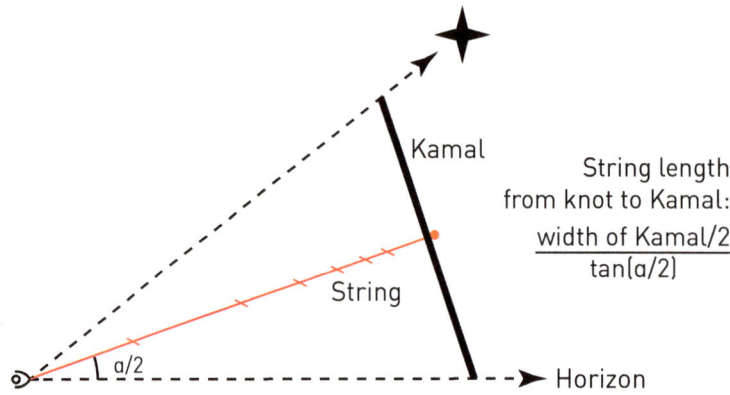

Figure 1: Principle of Kamal.

Origins

> The Kamal originated with Arab navigators of the late 9th century, and was employed in the Indian Ocean from the 10th century. It was adopted by Indian navigators soon after, and then adopted by Chinese navigators some time before the 16th century. [McGrail, 2004]

Name

> [...] requires in addition the knowledge of the measurement of stellar altitudes with the aid of the instrument known to most nautical historians as a *kamiil* — a word which never occurs in the Arabic texts. [Tibbetts, 1969].

General Construction

Figure 2: Kamal reconstruction. Includes the string with harmonically spaced knots. [Wndsn]

The Kamal consists of a rectangular piece of wood or horn ranging from about 1 × 2 inches up to 2 × 3 inches, to which a string with several knots, either equally or spaced in increasing increments, is attached through a hole in the middle of the card. (See figure 2 on page 3.)

The Kamal is used by placing one end of the string between the teeth while the other end is held away from the body roughly parallel to the ground. The device is then moved along the string, positioned so the lower edge is even with the horizon, and the upper edge is occluding a target star, usually Polaris because its angle to the horizon does not change with longitude or time. The angle can then be measured by counting the number of knots from the teeth to the device, or a particular knot can be tied into the string if traveling to a known latitude. Compare [Ferrand, 1928, Varadarajan, 2006].

The exact diameter of the Kamal isn't as important as the calculated knots for the specific diameter.

Wndsn XPD

As a rule, the smaller the diameter, the lower the maximum altitude measurable and the lower the resolution, due to bigger steps between knots.

In other words, to achieve the same range with a smaller diameter, the resolution needs to be lowered by way of increasing the degree step. (Compare table 2 on page 11.)

The closer the Kamal is held to the eye, the higher the altitude (maximum degrees) measurable. From this follows, that the larger the Kamal, the more maximum degrees are measurable. With a constraint about minimum knot distance added (see below), this means that a smaller Kamal, in order to achieve a higher number of maximum degrees measurable, has to have a lower resolution, due to the necessarily larger knot distances.

Knot-Spacing Algorithms

The knots were typically tied to measure angles of one finger-width. When held at arm's length, the width of a finger measures an angle that remains fairly similar from person to person. This was widely used (and still is today) for rough angle measurements, an angle known as *issabah* in Arabic. By modern measure, this is about $\frac{8}{5}$ degrees; or 1 degree, 36 minutes, and 25 seconds, or just over $1.6°$. It is equal to the arcsine of the ratio of the width of the finger to the length of the arm. With a fixed-width device, to measure equal increments (of one *issabah* in this case), the knots have to be spaced out across the string in a non-linear way, that means that the graduation interval is increasing in one direction for a tangential division. Compare [Raju, 2007, Mathew, 2018].

The formula for the derivation of the scales

The formula
$$\frac{Kamal\ side\ \times\ 5}{knot\ no.\ \times\ 6}$$
cited by [Prinsep, 1928] as well as by [Seydi Ali Reis] describes a graduation of $2°$ steps in altitude.

Wndsn XPD

> We take as a unit 5 times the diameter or side of the kamal and this length was divided into twelve parts; the first node is marked at the distance of 6 of these parts (counting from the kamal), and is called No. 12. Then the unit is divided into 11 and we take 6 of these new parts that are worn on the rope, and the point is called No. 11. The unit is successively divided into 10, 9, 8, 7 and 6 parts; when the marked knot coincides exactly with the length of 5 diameters, this point is numbered 6. A diameter beyond, gives division 5; one and a half beyond, gives division 4, which usually ends the ladder. [Prinsep, 1928]

And:

> Seydi Ali Reis relates the following instructions for tying the knots on the rope passing through the middle of the plate: First of all divide the gez into 12. By starting to count from the plate, tie a knot on the 6th point. This knot represents the beginning of the course, and it is equal to 12 *issabah* (finger breadth). This is the closest knot to the plate. Then the subsequent knots are tied. To find the place corresponding to 11 *issabah*, divide the gez into 11, and tie a knot on the 6th point. This is the second knot. The operation of tying knots on the rope goes on until the 7th knot is tied. This last knot corresponds to 6 *issabah*. [Danisan Polat, 2017]

Changing that formula to

$$\frac{Kamal\ side\ \times\ 6}{knot\ no.\ \times\ 6}$$

results in a ≈ 1.6° step which corresponds to the unit *issabah*.

> It is worth noting, that if the unit had been assumed at 6 diameters instead of 5, there would be a series of divisions almost identical with the *issabah* of 1° 36' used by the navigators of the fifteenth century. The series may also be extended both ways without very much deviating from the same progression: thus, commencing with tangential divisions, equivalent to the *issabah* from zero or sixteen *issabah*, or up to an altitude of 25°. [Prinsep, 1928]

Wndsn XPD

A simplified modern version of that formula, starting with variable degree values would be

$$\frac{Kamal\ side}{knot\ no.\ \times\ degrees}$$

(Note that $knot\ no.\ \times\ degrees$ is in radians.)

(Compare table 1 on page 8.)

The Cross-Staff, a Close Relative

An instrument related to the Kamal is the cross-staff, also used to determine angles, for instance the angle between the horizon and Polaris or the sun to determine a vessel's latitude, or the angle between the top and bottom of an object to determine the distance to said object if its height is known, or the height of the object if its distance is known, or the horizontal angle between two visible locations to determine one's point on a map. With the cross-staff, the string is replaced by a stick, the general principle remains the same. (See figure 3 on page 7.)

> The rules for dividing the wooden bar are the same as for the string, but the marks must be laid off invertedly, beginning at the eye end, which is in this the fixed point. [Prinsep, 1928]

Where with the Kamal, the string is graduated with knots starting from the Kamal plate and increasing harmonically towards the eye, towards higher degrees, with the cross-staff, the rigid staff is graduated increasing harmonically from the eye away, towards smaller degrees. The reason is that with the Kamal, we are moving the plate towards the eye, thereby shortening the string, while with the cross-staff, we are moving the cross-piece alongside the fixed-length stick.

Wndsn XPD

Description of the Instrument

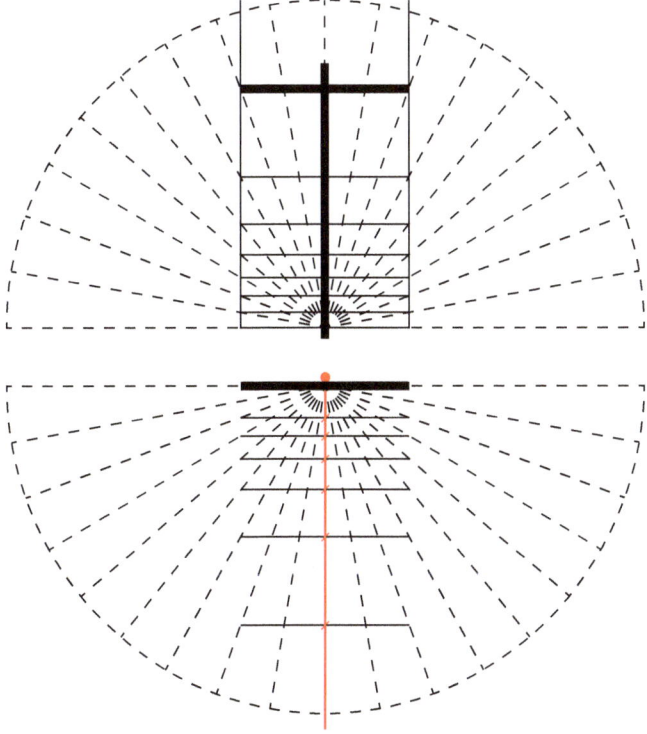

Figure 3: Kamal (bottom) vs. cross-staff (top) graduation. Derivation of knot-spacing: Kamal knots are calculated starting at the Kamal while for the cross-staff, the scales are calculated starting from the eye.

Table 1: Comparison of different knot intervals. Kamal side: 90 mm. Simplified (small angle subtension) formula for altitude: $arcsine(\frac{Kamal}{knot})$. Note that the step, the interval between knots is equal to the angle used in constructing the knot distances.

Variable Formula	1° 36' $\frac{Kamal}{knot \times 1\ issabah}$			5 $\frac{Kamal \times 5}{knot \times 6}$			6 $\frac{Kamal \times 6}{knot \times 6}$		
Knot	mm	in	Alt	mm	in	Alt	mm	in	Alt
4	802	31.6	6.44°	675	26.6	7.66°	810	31.9	6.38°
5	642	25.3	8.06°	540	21.3	9.59°	648	25.5	7.98°
6	535	21.1	9.69°	450	17.7	11.54°	540	21.3	9.59°
7	458	18.0	11.32°	386	15.2	13.49°	463	18.2	11.21°
8	401	15.8	12.97°	338	13.3	15.47°	405	15.9	12.84°
9	357	14.0	14.62°	300	11.8	17.46°	360	14.2	14.48°
10	321	12.6	16.29°	270	10.6	19.47°	324	12.8	16.13°
11	292	11.5	17.97°	245	9.7	21.51°	295	11.6	17.79°
12	267	10.5	19.67°	225	8.9	23.58°	270	10.6	19.47°
Step			1.61°			1.99°			1.64°

Wndsn XPD

Reconstructing the Kamal

Data

Based on the accounts reviewed in the literature, we have a few data points about the Kamal, more or less specific:

- Size: about 1 × 2 inches or more (which are nice and round numbers in the imperial unit system)
- Knot derivation: various formulas are known; see above
- String: some descriptions cite the use of 2 strings per Kamal

Questions

Using the above data, we can model a Kamal and attempt to answer the questions raised by the descriptions:

- Why the convenient inch measurements?
- What interval between knots?
- Why 2 strings per Kamal?

Size and Proportion Considerations

Identifying the variables

The size of the Kamal is determined by a number of variables: arm length, closest distance to focus, accuracy of knots and distance in between, minimum and maximum desired altitude to be measured, as well as the required precision of measurement.

As for precision, the first variable we have is the step between angles measured, e.g. 1 degree, 1 *issabah*, 2 degrees. The second variable is the minimum distance between knots on the string.

Inserting some numbers, we arrive at a size range that is suitable and ultimately we will see that it will be less important the exact size, as long as the knots are calculated on the size we have.

Determining the constraints

- Max. arm length: about 60 cm
- Closest distance to focus: about 10 cm[1]
- Minimum distance in between knots: about 10 mm (for practical knot-making reasons)
- Lower altitude: about $\leq 2.5°$
- Upper altitude: about $\geq 20°$
- Precision: $\leq 0.5°$ on the low end to $\leq 2°$ on the high end

Testing sample values for maximum altitude

Under these conditions, we can investigate the required minimum width of the long side of the Kamal to reach $20°$, depending on the step interval used:

---

[1] What's the closest distance humas can focus? The formula is $f = \frac{1}{D}$ where f is the focal distance in meters and D is the power of the lens in dioptres. The amplitude of accommodation is about 15 to 20 dioptres in the very young, decreasing to about 10 dioptres at age 25, and to around 1 dioptre above age 50. [Source: `https://www.aao.org/munnerlyn-laser-surgery-center/optical-properties-of-eye`]

Wndsn XPD

- The smallest side for 1° steps to reach 20° in altitude is 64 mm, starting at 6°.
- The smallest side for 1.61° steps to reach 20.9° in altitude is 48 mm, starting at 4.8°.
- The smallest side for 2° steps to reach 20° in altitude is 36 mm, starting at 4°.

(Compare table 2 on page 11.)

Table 2: Sample degree intervals and the resulting Kamal sides, low and high altitudes, as well as knots. (Method used: testing of various Kamal lengths and calculation of the resulting min. and max. values until requirements listed were satisfied.)

degrees	1.00°	1.60°	2.00°	
width	64	48	36	mm
from	6.00°	4.80°	4.00°	
to	20.00°	20.90°	20.00°	
knots	15	11	9	
from	181	130	102	mm
to	611	570	515	mm

Testing sample values for minimum altitude

A different approach is to determine the length of the short side of the Kamal based on the minimum degrees for the low altitude to be measured as well as the maximum arm length, then, in a subsequent step, determine the step necessary to reach the required high altitude.

(See *Calculating Optimum Kamal Size* on page 18.)

A Theory on Re-using Knots to Increase Precision

The Precision Trick

We have shown that, with the limitations of arm reach, minimum distance to eye, as well as minimum distance between knots, the instrument has a natural range and resulting size (or vice versa). Furthermore, the instrument's size is dependent on the required precision, if we know that value, the degree interval we need to graduate for, (together with the other requirements) the size of the Kamal is practically set.

So far, we only calculate with one side of the Kamal, and we arbitrarily chose the long side to derive the distances, knots, etc. Now, if we take a look at the table of knot distances, based on the precision (the degree interval) and calculate for both the long and the short side of the kamal, we can see some interesting relationships. (Compare table 3 on page 14.)

Hypothesis: use the string graduated for the long side, for the short side of the Kamal

If the sides have a relation of $\frac{1}{2}$, and we have a step of x degrees based on the long side, then we can use the same scale for the short

Table 3: 1 × 2 inch Kamal with string graduated in 1° steps with hints at corresponding values across scales and sides.

knot	step 1.00° angle	$cot = \frac{1}{tan}(\frac{angle}{2})$	Kamal in mm 50.8 $\frac{cot \times Kamal}{2}$	25.4	angle/2
1	1.00°	114.6	2911	**1455**	0.50°
2	2.00°	57.3	**1455**	727	1.00°
3	3.00°	38.2	970	485	1.50°
4	4.00°	28.6	**727**	363	2.00°
5	5.00°	22.9	582	**290**	2.50°
6	6.00°	19.1	**485**	242	3.00°
7	7.00°	16.3	415	208	3.50°
8	8.00°	14.3	**363**	182	4.00°
9	9.00°	12.7	323	161	4.50°
10	10.00°	11.4	**290**	145	5.00°
11	11.00°	10.4	264	132	5.50°
12	12.00°	9.5	**242**	121	6.00°

side and read $\frac{x}{2}$ degree steps. This means that for 1° steps on the long side, we can achieve a precision of 0.5° on the short side of the Kamal.

Using 1° as the interval has the advantage that the values read on the Kamal; the knots, are equal to the altitude to be measured.

To investigate that, we chose as the small side of the Kamal half the value of the long side. This means that, given the same formula of deriving the knot interval, we graduate the scale of the same length as the one for the larger side, but this time for a diameter half the size. This in turn, means that the degree interval for the smaller side is twice as large (since the proportion is $\frac{1}{2}$) as the one for the long side. If we arrange the scales by spacing out the values for the smaller side, we can clearly see the twice as large interval. But we can also see, that the scale for the larger side can fill in the gaps! (Compare table 4.)

This means that we can:

- measure from 18° to 10° in 1° steps on the 2 inch side

Wndsn XPD

- and from 9° to 2.5° in 0.5° steps on the 1 inch side (table 4 on page 15).

Table 4: Demonstrating the re-use of the scale for larger side on the small side for twice the precision (1° to 0.5°). (The lines mark the constraints.)

knot	angle	Kamal in mm		knot	angle/2
		50.8	25.4		
1	1.0°	2911		0.5	0.5°
2	2.0°	1455	1455	1	1.0°
3	3.0°	970		1.5	1.5°
4	4.0°	727	728	2	2.0°
5	5.0°	582		2.5	2.5°
6	6.0°	485	485	3	3.0°
7	7.0°	415		3.5	3.5°
8	8.0°	363	364	4	4.0°
9	9.0°	323		4.5	4.5°
10	10.0°	290	291	5	5.0°
11	11.0°	264		5.5	5.5°
12	12.0°	242	242	6	6.0°
13	13.0°	223		6.5	6.5°
14	14.0°	207	208	7	7.0°
15	15.0°	193		7.5	7.5°
16	16.0°	181	182	8	8.0°
17	17.0°	170		8.5	8.5°
18	18.0°	160	161	9	9.0°
19	19.0°	152		9.5	9.5°
20	20.0°	144	145	10	10.0°

So if we now use the scale of the larger side to measure altitudes with the smaller side, we can double our precision by being able to accurately measure half-steps, which wouldn't be trivial to eyeball with our non-linear, tangential graduation. Note that if we graduate a string for 2°, the values are the same as for the 1° scale, except that every other value is skipped and hence the knot distance is larger, increasing the range; by fulfilling our requirement to have a minimum distance of 10 mm between knots. (Compare table 5 on page 16.)

Wndsn XPD

Table 5: Demonstrating the re-use of the scale for larger side on the small side for twice the precision (2° to 1°). (The lines mark the constraints.) Note that the values are the same as for the 1° scale, except that every other value is skipped and hence the knot distance is larger.

		Kamal in mm			
knot	angle	50.8	25.4	knot	angle/2
1	2.0°	1455		0.5	1.0°
2	4.0°	727	727	1	2.0°
3	6.0°	485		1.5	3.0°
4	8.0°	363	364	2	4.0°
5	10.0°	290		2.5	5.0°
6	12.0°	242	242	3	6.0°
7	14.0°	207		3.5	7.0°
8	16.0°	181	182	4	8.0°
9	18.0°	160		4.5	9.0°
10	20.0°	144	145	5	10.0°
11	22.0°	131		5.5	11.0°
12	24.0°	119	121	6	12.0°
13	26.0°	110		6.5	13.0°
14	28.0°	102	103	7	14.0°

Two Strings?

Having one scale essentially divide the other in half steps, leads to the conclusion that we could have two strings for our sample Kamal, one graduated for 1° steps for the long side and to be re-used as 0.5° steps on the short side, as well as an additional string for the high range of 18° to 24° in 2° steps. (Compare figure 4 on page 17.)

The immediate question that comes up; why not combine the two strings into one, since most knots are the same, except for the high and low ends? The answer is usability; if we have just one string, we would have to combine steps of two resolutions, which may not be apparent in actual use and may lead to reading errors.

A way to mitigate the usability problem would be to mark the knots with different color strings threaded through each one of them. For example, by marking the 2° steps, the information necessary to use the instrument would be the degree value for the closest-to-the-Kamal knot as well as the step width. The mnemonic would be something like *"red in 2° steps down from 24° for the long side, and in 1° steps down from 12° for the short side; black are half steps."*

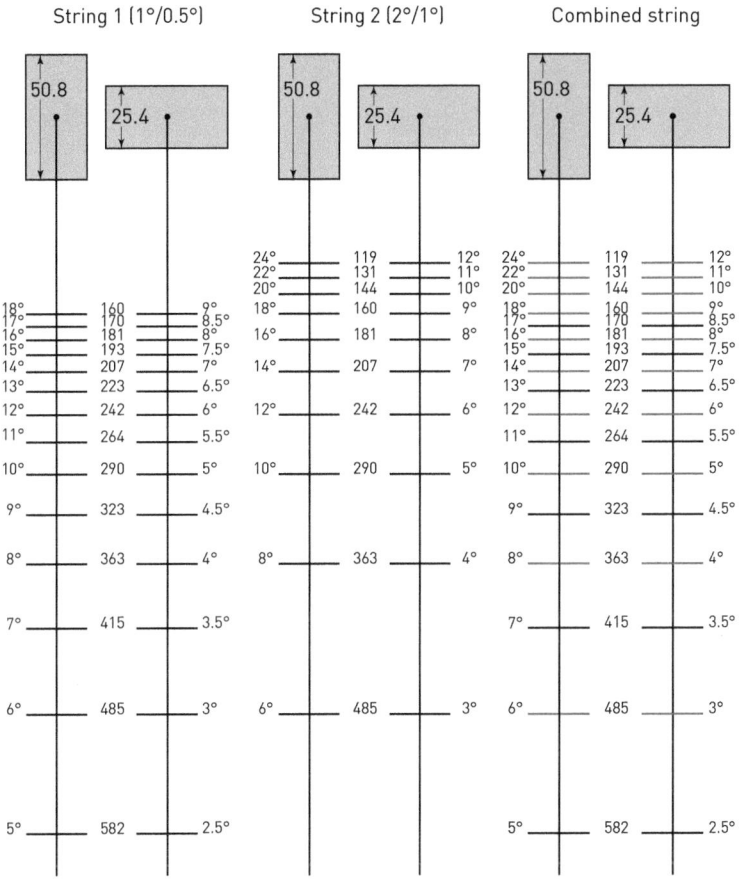

Figure 4: One string provides 1° and 0.5° and a second string provides 2° and 1° steps respectively. The string on the right combines all knots and marks those at 2° steps in red.

Calculating Optimum Kamal Size

For our constraints of distance to knot ≤ 600 mm and lowest angle ≤ 2.5 degrees, we can compute:

$$\frac{Kamal\ side}{rad(lowest\ angle)} = distance\ to\ knot$$

or

$$Kamal\ side = max\ distance\ to\ knot \times rad(lowest\ angle)$$

and we get 600 mm $\times$ rad($2.5°$) = 26.2 mm for the smaller side.

Graduating in $1°$ steps for the long side, and with our constraint of a minimum space between knots of 10 mm, we can measure in $0.5°$ steps from $2.5°$ to $9°$ with the small side. With the long side (26.2 mm $\times$ 2 = 52.4 mm), the maximum altitude measurable in $1°$ steps is $18°$. From $18°$ to $26°$, we can measure in $2°$ steps within the ≥ 10 mm between knots constraint.

Note that we set the long side to short side $\times$ 2, in order to be able to use the precision trick.

(Compare table 6 on page 20.)

The average optimal size of the Kamal accordingly results from the physical constraints of knot distance on the string, minimum eye-to-Kamal distance, maximum hand-to-eye distance, as well as the minimum and maximum degrees we intend to measure.

Dynamic resolution

Now that we have a string with different resolutions to be used for both sides of the Kamal, we note that the near resolution of $0.5°$ at $\leq 9°$ is 4 $\times$ as high as the far resolution of $2°$ at $\geq 18°$.

For an instrument measuring static distances, it makes sense to have the same resolution near and far, whereas for an instrument used in dynamic, moving navigation, an increase in resolution in the near field is preferable since we can approach with coarser values and close in with finer values.

Wndsn XPD

Conclusion

Putting together the data, our ideal Kamal has a size of 26.2 mm × 52.4 mm and one combined string with dynamic knot-spacing for the range of 2.5° to 26° altitude.

Answering the questions:

- Why the convenient inch measurements? The 1 × 2 inch Kamal size estimate seems to come close to our optimum size and may be the result of rounding to imperial units.

- What interval between knots? A convenient interval is the unit used for the altitude; if we measure the altitude in degrees, graduating the Kamal in degrees prevents conversion errors and we can "read" results directly.

- Why 2 strings per Kamal? Either for usability reasons, or to graduate strings with different intervals, or to benefit from double precision through re-using graduations.

Additional insights

- A maximum resolution of 0.5° translates to sufficient precision for navigation at sea; especially for naked-eye navigation on a moving ship.

- A dynamic range of resolution with higher precision in the near field is suitable for navigation while closing in on measuring targets with increasing resolution.

Wndsn XPD

Table 6: Knot-spacing for optimum Kamal dimensions.

| | Kamal in mm | | | |
| | 52.4 | 26.2 | | |
knot	angle	angle/2	mm	notes
1	1.0°	0.5°	3002	
2	2.0°	1.0°	1501	
3	3.0°	1.5°	1001	
4	4.0°	2.0°	751	
5	5.0°	2.5°	**600**	max. arm dist.
6	6.0°	3.0°	**500**	
7	7.0°	3.5°	**429**	
8	8.0°	4.0°	**375**	
9	9.0°	4.5°	**334**	
10	10.0°	5.0°	**300**	
11	11.0°	5.5°	**273**	
12	12.0°	6.0°	**250**	
13	13.0°	6.5°	**231**	
14	14.0°	7.0°	**214**	
15	15.0°	7.5°	**200**	
16	16.0°	8.0°	**188**	
17	17.0°	8.5°	**177**	
18	18.0°	9.0°	**167**	last knot w/ dist. $\geq$ 10 mm
19	19.0°	9.5°	158	skipped knot
20	20.0°	10.0°	**150**	from here in 2° steps
21	21.0°	10.5°	143	skipped knot
22	22.0°	11.0°	**136**	
23	23.0°	11.5°	131	skipped knot
24	24.0°	12.0°	**125**	
25	25.0°	12.5°	120	skipped knot
26	26.0°	13.0°	**115**	last knot
27	27.0°	13.5°	111	skipped knot
28	28.0°	14.0°	107	constraint reached

Wndsn XPD

Precision, Accuracy, and Range

Achieved Precision

Kamal precision is determined by how exact the knots are calculated in relation to the plates dimensions as well as how exact the knots are placed on the string. Further, string elasticity affects precision of the device.

Higher altitude *ranges,* i.e. the range of altitude covered by the instrument, mean bigger intervals, and hence a lower overall or partial measuring resolution. (Compare figure 4 on page 17.)

Measuring in the lower degree range, the smallest division we have determined (as per the knot-spacing) to be 0.5°.

Using the rule of precision:

> The maximum achievable precision is ± half of the smallest division of the scales. [ISO, 1995]

With that, 0.5° divisions result in a *maximum achievable precision* of 0.25° or 15' of latitude, which translates to 15 nautical miles.

So for our navigation, we have a circle of inaccuracy with a radius of 7.5 nautical miles (to see the horizon at a distance of 7.5 nautical miles, we'd have to be about 14 m above the water, or conversely, from sea level, we would be able to see structures taller than 14 m at a distance of 7.5 nautical miles).

Achieved Range

The maximum north-south range of the device can be calculated as:

26 - 2.5 = 23.5 × 60 = 1,410 nautical miles

with 60 nautical miles/degree and an angular range from 2.5° to 26° of measured altitude.

Also note that:

> Further the highest number, 12, gives nearly the latitude of Calcutta, or 22° 38', the most northerly latitude for which the Maldive navigators have any occasion; while the lowest mark, 4, gives the latitude (nearly) of the southern point of Ceylon, or the average of the Maldive islands. [Prinsep, 1928]

Accuracy

The accuracy of the Kamal is constrained by knot step and alignment to horizon influenced by visibility, mirage, etc. While moving between knots, accuracy is reduced until approaching the next knot, where accuracy increases again.

> Accuracy is determined and limited by the precision with which physical markings can be created and reproduced, as well as read and aligned to the corresponding object to be measured.

Accuracy is determined by the ability to read the scale marks, or with the Kamal, the ability to align the plate to the object to be measured.

While the device precision is dependent on the scales, the accuracy of reading is dependent on external factors, some of which are controllable to an extent.

Reading accuracy may be affected by light and atmospheric conditions like mirage around the object to be measured, as well as factors that prevent steady scales for a reliable read.

Wndsn XPD

Bibliography

[Burch, 2008] Burch, D. (2008). *Emergency navigation: find your position and shape your course at sea even if your instruments fail.* International Marine / McGraw-Hill, Camden, Me.

[Congreve, 1850] Congreve, H. (1850). A brief notice of some contrivances practised by the native mariners of the coromandel coast in navigating, sailing, and repairing their vessels. *The Madras Journal of Literature and Science*, (XVI):103–104.

[Danisan Polat, 2017] Danisan Polat, G. (2017). Kamal, an instrument of celestial navigation in the indian ocean, as decribed by ottoman mariners piri reis and seydi ali reis. *Osmanli Bilimi Arastirmalari*, 19:1 – 12.

[de Saussure, 1928] de Saussure, L. (1928). Commentaire des instructions nautiques de ibn majid et sulayman al-mahri. In Ferrand, G., editor, *Instructions nautiques et routiers arabes et portugais des XVe et XVI siècles*, pages 129–175. Geuthner.

[Ferrand, 1928] Ferrand, G. (1928). *Instructions nautiques et routiers arabes et portugais des XVe et XVI siècles.* Inst. für Geschichte d. Arab.-Islam. Wiss. an d. Johann Wolfgang Goethe-Univ., Frankfurt am Main.

[ISO, 1995] ISO, editor (1995). *Guide to the expression of uncertainty in measurement: (GUM).* International Organization for Standardization, Geneva.

[Mathew, 2018] Mathew, K. S. (2018). *Shipbuilding, navigation and the Portuguese in pre-modern India.* Routledge, Taylor et Francis Group, London and New York.

[McGrail, 2004] McGrail, S. (2004). *Boats of the World: From The Stone Age To Medieval Times.* Oxford University Press, Coimbra.

[Prinsep, 1928] Prinsep, J. (1928). Note on the nautical instruments of the arabs. In Ferrand, G., editor, *Instructions nautiques et routiers arabes et portugais des XVe et XVI siècles*, pages 1–24. Geuthner, Paris.

[Raju, 2007] Raju, C. K. (2007). *Cultural foundations of mathematics: the nature of mathematical proof and the transmission of the calculus from India to Europe in the 16th c. CE.* Pearson Longman, New Delhi.

[Tibbetts, 1969] Tibbetts, G. R. (1969). *The navigational theory of the arabs in the fifteenth and sixteenth centuries.* Junta de Investigacoes do Ultramar, Coimbra.

[Varadarajan, 2006] Varadarajan, L. (2006). *Indo-Portuguese encounters: journeys in science, technology, and culture.* Indian National Science Academy, New Delhi.

About Wndsn

Wndsn's Applied Science Lab, based in Berlin, develops and manufactures that which can't be improvised; measurement, navigation, and surveying instruments informed by the motto *"Ex Mensura, Scientia"* — knowledge from measurement.

Wndsn produces archival quality products that are designed with intent by combining techniques proven over centuries; arcane science meets cutting edge contemporary methods, resulting in iconic, timeless, high-utility designs.

In addition to custom-built instruments and tools, metrology & illumination solutions, Wndsn creates expedition mementos and morale patches to celebrate cross-disciplinary exploration in the spirit of the Renaissance. Wndsn morale patches are acutely designed — no line is left to randomness, no element is mere filler. They serve as infographics, how-tos for the Wndsn tools, magic sigils, as well as functional markers.

Contact

For feedback, inquiries, suggestions, or custom Telemeter nomographs, please feel free to contact us at:

- info@wndsn.com

Links

- Wndsn Navigation Tools: https://store.wndsn.com/
- High-precision Range Calculator: http://tycho.wndsn.com/

www.ingramcontent.com/pod-product-compliance
Lightning Source LLC
Chambersburg PA
CBHW040348220526
45473CB00009B/2814